© Aladdin Books Ltd 2001

*Designed and produced by*
Aladdin Books Ltd
28 Percy Street
London W1P 0LD

*First published in
the United States in 2001 by*
Copper Beech Books,
an imprint of
The Millbrook Press
2 Old New Milford Road
Brookfield, Connecticut 06804

ISBN 0-7613-2464-X (lib. bdg.)
ISBN 0-7613-2334-1 (paper ed.)

*Cataloging-in-Publication data is on
file at the Library of Congress*

*Printed in Belgium
All rights reserved*

*Coordinator*
Jim Pipe

*Design*
Flick, Book Design and Graphics

*Picture Research*
Brian Hunter Smart

*Illustration*
Mary Lonsdale for SGA

*Picture Credits*
Abbreviations: t – top, m – middle,
b – bottom, r – right, l – left, c – center.
All photographs supplied by Select
Pictures except for: Cover, 10tl, 17,
20-21, 23br — Corbis. 3, 4-5, 6tl, 18-19,
20tl, 22tr, 22bl, 24tr, 24ml — Digital Stock.
4tl, 8tl, 22tl — John Foxx Images.
9 — Danny Lehman/CORBIS. 12tl,
23ml — Stockbyte. 15 — Wolfgang
Kaehler/CORBIS. 18tl — CORBIS.

My World

# Slow and Fast

By Dr. Alvin Granowsky

Copper Beech Books
Brookfield, Connecticut

# Fast

Kate and Dan are going to the fun park.

The car goes fast along the road. Kate and Dan can't wait to get to the park.

What else is fast?

A cheetah is fast.

It runs as fast as a car!

# Slow

What is slow?

A traffic jam is slow.
Kate and Dan want to get to the park, but the cars crawl along the road.

A snail is slow, too.
It crawls along the ground.

# Faster

They arrive at the fun park.
The children run to the slides.

Kate runs faster than Dan.
But Dan is faster than
Mom's friend Julie.

Who is faster?

Dan sits on a mat.

He slides faster than Kate.

He gets to the bottom first.

 # Slower

Who is slower?

Dan is slower. Kate says he is as slow as a turtle. But when Kate runs too fast, she falls down.

Sometimes slower is better.

The man on the rope is slow.

If he goes faster, he will fall off!

# Going faster

What speeds up?

This ride speeds up!
It goes faster and faster.

Then the ride slows down.

It goes slower and slower.

When it stops, Dan and Kate get off.

# Fastest

The go-karts go fast.

Mom's kart is the fastest.

It goes like a rocket!

Racing cars are very fast.

They race each other.

The fastest car wins the race.

# Slowest

Who is slowest?

Julie is the slowest eater.
She takes a long time!

A sloth is one of the slowest animals. It can take all day to walk along a branch.

# Spinning

What spins fast?

The seats in this ride spin fast.
They go around and around.
Dan and Kate get dizzy!

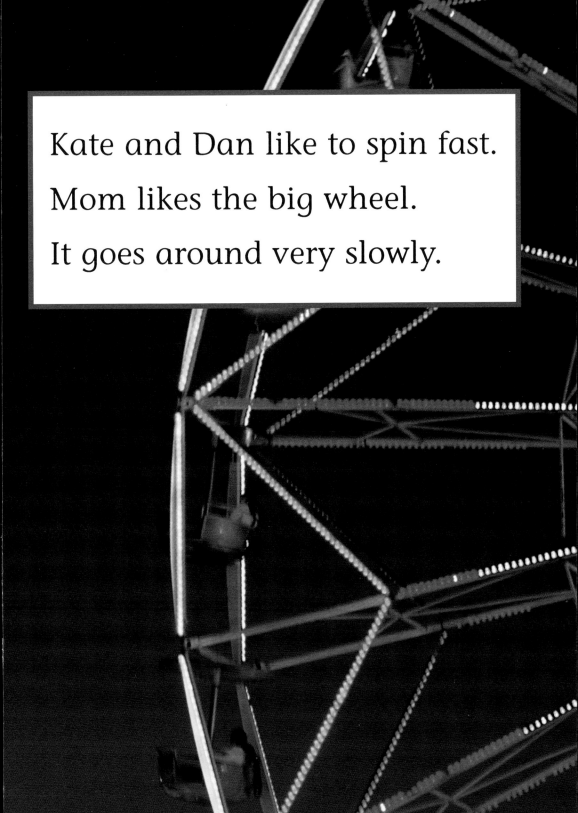

Kate and Dan like to spin fast.

Mom likes the big wheel.

It goes around very slowly.

# Falling

Kate and Dan watch the air show.

A sky diver jumps from a plane. He falls faster and faster.

When the parachute opens, the sky diver slows down.

# Slow and fast

What is slow and fast?

A roller coaster goes up the hill slowly.
But it comes down fast!

Everybody screams!

# Here are some words about speed.

Slow      Fast

Slower   Faster                    Slowest

Fastest

Speed up

Slow down

# Are these things fast or slow?

Ride

Car

Rocket

Bicycle

Roller coaster

Can you write a story with these words?

*Do you know?*

A watch times how fast we can run.

A dial tells us how fast a car is going.

Signs tell people not to drive too fast.